Making
Science Work
For All

Making Science Work For All

Can Participatory Budgeting Schemes Be Applied to Scientific Discoveries?

ADRIAN SOH

Royal Melbourne Institute of Technology (2016)

PARTRIDGE

ISBN: Softcover   978-1-4828-6426-7
             eBook   978-1-4828-6427-4

Print information available on the last page.

To order additional copies of this book, contact
Toll Free 800 101 2657 (Singapore)
Toll Free 1 800 81 7340 (Malaysia)
orders.singapore@partridgepublishing.com

www.partridgepublishing.com/singapore

ACKNOWLEDGEMENTS

I wish to acknowledge that many people have contributed to creating this book and that it has become a better book because of their comments and ideas. I wish to thank Professor John King, Professor Paul Fisher, Associate Professor David Prentice, Dr Joseph Vecci, Anthony Rossiter, Matthew Cary, and Rick Garotti.

INTRODUCTION

This book considers whether building a civil society in a country around the public funding of science would simultaneously achieve greater economic equality through a participatory budgeting scheme for science. A civil society or culture built around the public funding of science is one in which there is a national culture or civil society that accepts and is willing to defend the public funding of science as a national priority over the long term, even over other government spending priorities. In a civil society and culture of science, a large portion of the national labour force participates and believes they benefit directly from a large-scale funding of science and technology.

A participatory budgeting scheme for science would give money first to technology start-ups and then to pure research. It would give a country's tax-paying citizens a financial distribution every two years once the assets and returns of the fund have grown to a defined level. The fund would be run by a board independently of government departments and could be dissolved only by a majority vote of taxpayers who lodged tax returns within the intervening two-year period between financial distributions.

How can it be ensured that such a system would create groundbreaking (original or pioneering) scientific research? Furthermore, how would the funds be allocated to technology start-ups and pure research (which is research that improves scientific theories for improved understanding or prediction of natural and other phenomena)? How would dividends to citizens be distributed? What would be the frequency of the fund's distribution? How would a participatory budgeting scheme manage risks? How would it manage to diversify pure research? How would it assist joint domestic and foreign research projects? These are the questions that this book seeks to address.

Research and a number of reports and have suggested that overall, there are significant benefits to be gained from science funding. The Organisation of Economic Cooperation and Development (OECD) argues in various reports that consistent and strong science funding can boost the growth and productivity of economies (OECD 2013: 56). The Centre for International Economics stated that science and technology spending would lead to a reduction in the price of a wide variety of goods and would create savings in terms of money and resources that could be returned to the economy for better use. Though there would be a twenty-year period between initial investments and discoveries, there would also be a significant spillover effect in multiple sectors of the economy, including banking, mining, and telecommunications.

This spending would also preserve a workforce that would translate research results from overseas and would enable those technologies to be integrated into the Australian economy. Investments in core subjects such as mathematics, science, physics, and chemistry allowed for the development of new disciplines such as earth sciences (Centre for International Economics 2015: 5–7). Philip Smith, a former executive officer of the National Academy of Sciences (NAS) and National Research Council (NRC), and Michael McGeary, a former senior staff officer and study director at NAS and NRC, agree with the OECD that increased funding to science brings productivity to economies and the discovery of new industries. In their view, science spending renews science infrastructure, and any cuts in spending must be selective (Smith and McGeary 1997: 37). Andrew Moore of the European Molecular Biology Organisation argues that the public funding of health science results in great savings to public health systems (2000: 330). Therefore, maintaining a steady growth in funding of science is crucial. Sections 1 and 2 show how a participatory budgeting scheme for science would help in this aim.

Furthermore, writers Mitcham and Frodeman stated that the plateau in science funding led to a reduction in the number of science and engineering graduates as career opportunities dried up (2002: 86). Howard and Laird agreed with Mitcham and Frodeman's argument about skills. They argued that it is necessary to attract and train younger scientists to renew research, reduce labour bottlenecks in science research, and allow for the replacement of older scientists who eventually retire. If these early career scientists were to be lost to other sectors of the economy, it would be hard for them to re-enter the field of scientific research (Howard and Laird 2013: 72). A 2012 report to then Australian Prime Minister Julia Gillard by Australia's chief scientist, Professor Ian Chubb (Chubb, Findlay, Du, Burmester, and Kusa 2012: 7–10) recommended incentives (such as increased scholarships and reduced student debt) to attract human resources, including women, to science and mathematics. They also recommended the setting up of regional science centres. The participatory budgeting scheme for science could grant funds based on the number of personnel trained, as stated earlier, as the scheme's assets and returns surpass a certain level. Therefore, due to this difficulty in gaining back scientific labour lost, sections 1 and 2 show how a participatory budgeting scheme for science would assist in maintaining a steady growing source of public funding for science. It would therefore also assist in maintaining jobs in the science and technology sector of countries that adopt it.

However, public science spending continues to be an easy target for restraint in government spending. Appendix 2 shows that Australian and OECD government spending on science has oscillated with economic cycles rather than being held steady over time. However, this oscillating public spending on science and technology disrupts research and planning of research priorities. Therefore, a steady real increase in public funding for science and technology is more ideal (see Appendix 3).

In taking into account the affect of inflation on Australian government spending on science, spending has seen a real cut (see Appendix 2). Australia is not alone in this habit of oscillating public funding for science. Many countries of the OECD cut public science spending in the midst of an economic downturn when they should have increased it to enable better productivity when economic recoveries took place (see Appendixes 4, 5, and 8).

Howard and Laird argued that the allocation of a significant amount of funds to a relatively small number of people was an easy target for fiscal restraint (2013: 71–76). For example, Ian Chubb, Australia's chief scientist under the Gillard government in Australia (2010–13) pointed out that Australia produced only 200,000 graduates of science and mathematics per decade and only fifteen out of 1000 Australians worked in the science and technology field (Chubb, Findlay, Du, Burmester, Kusa 2012: 16). Overall, by world standards, newer science countries like Australia remain behind other new countries of science such as South Korea and Israel and older countries of science such as Germany, Japan, the United States, and France. The conclusions are based on public funding of science as percentages of GDP and on researchers per thousand employed.

Australian research and development expenditures also remain small when compared with many countries, such as South Korea, Japan, Germany, the United States, and China (see Appendixes 1 and 4). Australia's spending on research and development spending is respectable as a percentage of GDP and on researchers per thousand people in the labour force. However, when compared to the absolute money spent and human resources devoted to research and development by other countries, Australia ought to be spending more as a percentage of GDP to compete with scientific research by the likes of China, South Korea, the United Kingdom, the United States, and Germany (see Appendix 4, 6, and 7).

Furthermore, countries like China and South Korea have consistently increased their shares of world research and development. At this rate they would catch up or surpass that spent by other countries. Therefore, maintaining a steady growth in funding of science is crucial. Sections 1 and 2 show how a participatory budgeting scheme for science would help in this aim of growing a science culture and civil societies through giving the public direct financial benefit from scientific commercialisation and discoveries.

Science spending also loses out when competing against more pressing spending priorities. The Australian National University (ANU) Public Opinion on Science found that of 1200 respondents who were asked "What do you think is the most important problem facing Australia today?" only 0.7 per cent said infrastructure, planning, and innovation. This was in contrast to 16.9 per cent who answered the economy and jobs, 15.3 per cent environment and global warming, and 13.8 per cent immigration (Lamberts, Grant and Martin 2010: 19–20). Sections 1 and 2 show how a participatory budgeting scheme for science would help in this aim of growing a science culture and civil societies through giving the public direct financial benefit from scientific commercialisation and discoveries.

Foltz argued that participation and science awareness are heightened only when the public has a direct stake in certain issues such as medical science. Equally, policymakers were convinced that the public is incapable of understanding the intricacies of science policy (Foltz 1999: 118). Howard and Laird agreed that the public's misunderstanding of science disqualifies them from making competent decisions about the allocation of science funding (2013, 71–76). For various reasons it appears the public is disconnected with public science funding, again making it a popular target for budget cuts when spending ought to be maintained. Sections 1 and 2 show how a participatory budgeting scheme for science would help in this aim of growing a science culture and civil societies through giving the public direct financial benefit from scientific commercialisation and discoveries.

Foltz argued that a "technocratic decision-making system" of powerful experts made all the science spending decisions for the public, but lacked sufficient legitimacy with the public to have sole control over the allocation of public resources even though the public is most affected by the science produced, and their taxes funds a large amount of scientific research (1999: 117–19). Sections 1 and 2 show how a participatory budgeting scheme for science would help in allowing the "technocratic decision-making system" gain legitimacy with the general public and better allocate resources. This would be achieved through giving the public direct financial benefit from scientific commercialisation and discoveries, amending grant criteria and hiring external consultants to advise it.

Section 1 discusses the concept of participatory budgeting for science. Section 2 outlines how a participatory budgeting scheme for science would operate and how the risks would be mitigated. Section 3 details the state of science funding in the world today. Section 4 discusses arguments for and against the public funding of science and how the participatory budgeting scheme for science would counter the problems in the current public funding model for science. Section 5 concludes the arguments presented and outlines areas of further investigation and the implementation of such a scheme.

SECTION 1

What Is the Participatory Budgeting Scheme for Science?

A participatory budgeting scheme for science would be similar to one envisaged by the British economist James Meade. He advocated for the state to set up an investment fund that owns a portfolio of assets. Its returns would be distributed to citizens in the form of a social dividend (White 2014: 67–69). This state investment fund is similar to the participatory budgeting scheme for science. However, the scheme would be totally devoted to advancing science discoveries and their commercialisation. Dissolving it would require a vote of stakeholders, to reduce the likelihood of a change in the scheme's funding levels with a change of government. Meade, author of the book *Efficiency, Equality and the Ownership of Property* (1964), wrote about four scenarios that could possibly occur with the automation of production and the gradual depression of wages as businesses try to maintain their competitive advantage (White 2014: 67–69).

The welfare state sought to solve income inequality through increased tax and transfers from the owners of capital to waged labour. Meade was sceptical of this approach because it reduced the incentive to work (White 2014: 67–69). Meade argued that a trade-union state is one in which the state that creates labour legislation to maintain the minimum wage and strengthen the ability of trade unions to collectively bargain on behalf of workers would cost jobs by increasing the cost of production. It also would erode a country's competitiveness (White 2014: 67–69).

Post-Keynesian economist Philip Whyman said that a trade-union state would create "conflict inflation". This occurs when a wage increase has been won, after which business owners increase prices in order to restore their profit margins, thereby sparking another demand for increased wages by unions (Whyman 2006: 233). Meade thought the threat of bankruptcy by firms limits the ability of government and trade unions to redistribute wealth through taxation and transfers or industrial agitation (Whyman 2006: 233). These limits were seen by Swedish social democratic economists Gosta Rehn (1957: 101–3) and Rudolf Meidner (1981: 308) as well. Meade was also opposed to the nationalisation of previously privately-owned firms due to the economic disruption it would cause to private-sector investment. It was because of the problems of each of the above scenarios or states that Meade proposed his government investment fund that would pay a social dividend to citizens.

According to Whyman, British economist John Maynard Keynes also argued for governments to facilitate investment and ownership of property for lower-income earners. This would allow for greater and more sustained consumption and investment in the economy, which in turn would curb the roughness of the occasional economic uncertainty and speculative tendencies of higher-income earners (2006: 229–30). Donald George concurred that Keynes had been the original proponent for a state-sponsored socialisation of investment (1993: 470). The socialisation of the public investment in science through a participatory budgeting scheme would have the same positive economic effects envisioned by Meade and Keynes. It would continue to supplement incomes, reduce conflict inflation, and feed productivity and innovation. This is what would happen under the participatory budgeting scheme for science, as explained in section 2.

SECTION 2

Operation and Risk Management of a Participatory Budgeting Scheme for Science

As an example of how a participatory budgeting scheme would function, consider the following example. Between 2007 and 2016 (see Appendix 2), $77,726,200,000 AUD was spent or budgeted to be spent on various programmes in science and technology in Australia (Ministry of Industry and Science of Australia 2015). According to the Australian Treasury, 12,776,065 Australians lodged tax returns in the financial year 2014–15 (Australian Taxation Office 2015). Under a participatory budgeting scheme, all these taxpayers would be eligible to receive a dividend from the fund once it accumulates a certain amount of surplus funds.

According to a report commissioned by the chief scientist of Australia, the direct economic impact that science and technology spending had on the Australian economy is in the magnitude of $145 billion AUD per annum. Public investment in science was estimated to have created $75 billion AUD per annum in exports to the world (Centre for International Economics 2015: 44–47). If, for example, the fund then received $25 billion AUD in returns on its investments, then the tax lodgers would receive $1,956 AUD in extra income from the fund per annum. Some of the returns would have to be retained by the fund, however, in order to grow its portfolio of science and technology investments as well as perhaps to divert some funds to pure research at the university or institutional level when the returns and assets reach an appropriate level. However, the public-good aspect of the scheme (meaning that no one is excluded, and consuming its services would not reduce consumption by other businesses) means it would be difficult to estimate how much distribution would be gained by citizens who participate in the scheme. This is especially the case with constant changes to the numbers of taxpayers and revenues.

The government would retain some right to allocate the annual science budget according to national needs in order to avoid a problem of asymmetrical information (a situation where one party has superior information to another party in an economic transaction) that taxpayers would have when choosing what pure research to support.

Furthermore, perhaps governments would allocate only part of their science and technology budgets in such a participatory budgeting scheme in order for the scheme to focus on financing only commercialised technology. The government's budgets, in the interim, would focus on pure science research until the assets and return of the participatory budgeting scheme grows to a level where it could both return dividends to taxpayers directly and give grants for pure science research. The government would also make a contribution to the fund every year. The participatory budgeting for science scheme would make distributions to pure research and taxpayers every two years when assets and returns of the scheme grow to a certain level.

However, the purchase of equity in or lending to start-up firms and commercialised technologies would occur in alternate years as well. The participatory budgeting scheme would also have a board that is independent of the government. External auditors and experts would be hired to assist it with the problem of asymmetric information. In the case of the death of citizens who submit a tax return, the dividend of the deceased would be distributed amongst other citizens.

Furthermore, it should be difficult for a participatory budgeting scheme for science to be dismantled, even with a change in government. Governments should be forced to seek the approval of the stakeholders (all those who submit a tax return) for it to be dismantled.

In order to gain foreign science and technology knowledge, the remaining science budget and participatory budgeting scheme would be allowed to make grants, loans, or equity purchases in joint domestic and foreign research and commercialised technology. Equity (the value of ownership interest) stakes or lending by the participatory budgeting scheme in commercialised technologies would either be kept with current share registry firms in Australia or would be internally recorded by the scheme's bureaucracy at a comparable cost.

The participatory budgeting scheme and any fund that is assigned to manage funds runs the risk that technologies commercialised by the fund might not become as financially viable as first thought. However, strategies have been adopted by states all over the world to manage such risks. The Clean Energy Finance Corporation (CEFC) in Australia, for example, lends funds based on the credit rating of firms that seek its financing. In the CEFC, a highly experienced portfolio management team monitors all of its investments by examining how the portfolio performs against industry benchmarks and guidelines.

Furthermore, the CEFC is not allowed to acquire a majority equity ownership of any venture. All investments must involve only domestic firms. The minimum investment size is set at $20 million AUD. It cannot place more than 30 per cent of its portfolio in a single technology area (say biotechnology), industry, or geographic location. Investments are made in later-stage technology. The CEFC also does not invest in firms that have liabilities constituting more than 80 per cent of their equity.

The CEFC participates in pooled financing schemes and partnerships for small and medium businesses. It also has to ensure that their investments are audited by external auditors. External financial consultants are hired to advise on their investments. The equity investments of the CEFC are limited, while debt guarantees are more strongly encouraged (Clean Energy Finance Corporation 2015). The CEFC is severely limited in its ability to borrow funds from private banks to finance its operations. Furthermore, the CEFC has limits ($300 million AUD) on how much they are able to grant in terms of guarantees of loans made by the private financial institutions to clean-energy firms.

Loan guarantees are limited to only 5 per cent of the CEFC special fund, a separate fund that grants loan guarantees to banks that loan funds to the private sector.

Appointees to the CEFC board are required to have a background in such disciplines as venture capital, economics, or the law, and must not be employed by the Australian federal or state government. Any appointee to the board for the participatory budgeting scheme for science would have to have a background in science too.

The corporation is allowed to buy derivatives (contracts between buyer and seller entered into today regarding a transaction to be fulfilled at a future point in time) to safeguard the value of assets (Clean Energy Finance Corporation Act 2012). A participatory budgeting scheme for science would be able to employ these measures to limit the risk of investing in loss-making ventures. External auditors and experts would address problems of gathering information about what to invest in.

The CEFC is able to gain significant yields from its investment on behalf of taxpayers. In the financial year of 2013–14, its corporate loans yielded 7.8 per cent, project finance yielded 8.1 per cent, its co-financing programmes yielded 5.2 per cent, its equity investments yielded 8.5 per cent, and its portfolio yielded 7 per cent return for Australian taxpayers. The CEFC claims that it encourages private sector investment in the Australian renewable energy sector on a 1 to 2 basis, which means that for every dollar invested in projects they generated two dollars of private sector investment. According to the 2013–14 financial report of the CEFC, the corporation recorded no defaults, though many loans were due in five years from 2013 (CEFC 2014). Therefore, if the participatory budgeting scheme for science were to adopt similar risk strategies to the CEFC, the financial rewards could be replicated.

Venture capitalists also employ certain methods to limit risk to their portfolios in addition to investment diversification. These methods include the prescreening of investments to be made (such as pre-investment audits) and staged payments (not giving the firm in which they are investing all the money that they have pledged at once). These measures are designed to counteract the problem of asymmetric information, which is where one party in a transaction, such as an entrepreneur, has all or better information about it than another. This imbalance could lead to an unsuccessful investment by a fund manager.

Venture capitalists have used contracts that penalize the entrepreneur, such as to increase a fund manager's cost of compensation if the financial outcome is negative, or to cease further financing. Entrepreneurs could also be asked to co-finance their technology venture to reduce the risk of failure and to provide incentives to reach outcomes. For smaller firms, regulations of these types might weigh heavily upon them, and there might be higher transaction costs for higher portfolio diversification (Kut, Pramborg, and

Smolarski 2007: 53–56), hence the need for a pooled finance scheme supported by funds from the participatory budgeting scheme for science similar to that of the CEFC. These methods of risk diversification would be used in order to guard against the risk of financial loss, especially as the participatory budgeting scheme would hire financial experts to advise it on future and less risky investments.

Norway's sovereign wealth fund (which stores Norway's oil wealth, to be spent in the event of a decrease in the world price of oil, Norway's main export) restricted equity investments of the fund initially to 40 per cent of fund assets until 2007. In addition, the fund places limits on its portfolio, such as assets denominated in specific currencies. It cannot own more than 5 per cent of a single firm's shares. Further, it employs the services of external and internal portfolio managers and relies on international credit ratings when investing in fixed income assets or government bonds. It was found that the fund outperformed the world stock market (with a return of 0.39 per cent per month to 0.24 per cent on average for the world stock market), but had a higher risk profile than the world stock market (4.71 standard deviation in monthly returns to 4.21 per cent). The participatory budgeting scheme, like the Norwegian sovereign wealth fund, would publish detailed reports on revenue, portfolio, expenditure, and profit for the government and the public (Caner and Grennes 2010: 599–605). Therefore, all these regulations can be implemented by the central fund of the participatory budgeting scheme for science to limit the risk of loss of taxpayers' funds.

Besides the issue of a public fund or participatory budgeting scheme for science managing risk, there is also the question of how such a fund would treat large, small, and medium businesses differently when creating innovation through small and medium businesses. The fund could create limited amounts of loan guarantees to banks or directly invest, or loan funds based on different criteria than those of large firms. The limited amounts of loan guarantees and some collateral requirements would assist against the possibility of moral hazard, which is when insurance inspires reckless behaviour (ADB and OECD 2014: 91). Bank lending or loans from state funds tend to be necessary for small and medium businesses because of owner and control changes, lack of credit history, and difficulty of presenting sophisticated financial reports. The main standard at which credit could be given is asset-based finance, where lending takes place based on specific assets such as receivables, machinery, equipment, or real estate.

Other mechanisms to funnel investments to start-ups could include making investments subject to less stringent credit ratings or loaning lesser amounts of cash. Giving loans and investment based on the personal credit histories of the owners of the start-up could also be an option. In some countries, like Japan, credit for small and medium start-ups were directed to sectors that had exhibited high growth rates for a certain number of financial quarters (OECD 2015: 6–15; ADB and OECD 2014: 26, 143–4). The participatory budgeting scheme for science could create a certain number of small-and-medium businesses loans and investments and then securitise them, which happens when various types of contractual debt are pooled and sold to investors or banks. Governments would also be able to guarantee this small-and-medium business loan securitisation. (OECD 2015: 49–68, 143–4). All this would assist the situation that small-and-medium start-ups would otherwise face, which is higher interest rates, shorter loan maturities, and higher collateral requirements when businesses are forced to offer financiers assets in the case of default (ADB and OECD 2014: 16–20).

Further assistance to small and medium start-ups could be given in the form of lower or subsidised interest rates. Longer repayment times could also boost small and medium business once loans or investments have been made by the participatory budgeting scheme for science (OECD 2015: 49–68). The fund could also hire some of its financial specialist consultants to assist small and medium start-ups to produce financial plans and information. Start-ups that receive funding from banks and the participatory budgeting scheme for science could be subject to repayments based on profitability and revenue stream or depending on the success of the start-up (OECD 2015: 6–15; ADB and OECD 2014: 26).

Other lessons can be learnt from the example of the Swedish wage-earner funds, a successful form of socialised private investment. The founder of this plan, Meidner, argued for Swedish businesses to be taxed and the money placed into funds to buy worker shares in Swedish firms. Rehn and Meidner argued that wage-earner funds supplement the incomes of workers to compensate them for wage demand restraints and increase the collective ownership of capital, which is something that would be achieved under a participatory budgeting scheme for science.

After heavy resistance from the Swedish business sector (also from other Swedish left-wing players), a watered-down version of Swedish wage-earner funds legislation was enacted in 1983 (Pontusson and Kuruvilla 1992: 785). The Swedish government used excess profit taxes (20 per cent on pre-tax profits) and payroll taxes (0.2 per cent of a firm's payroll) to fund this tool of worker ownership through general taxation. The taxes were levied according to their profitability and the size of their payroll, which was profits of over 1 million Swedish kronor and 6 per cent of their total payrolls, whichever was higher when taking into account inflation and other appropriations. Financing these wage-earner funds was limited to a seven-year period following its passage in 1983.

No fund could own more than 8 per cent of any corporation or enterprise (which was later reduced to 6 per cent). Originally, it was intended that the funds' share purchasing would go on for perpetuity. However, when it was legislated, taxes to provide funds for the wage-earner funds were collected for only seven years (Pontusson and Kuruvill 1992: 784). Therefore, from the experience of the Swedish wage-earner funds, the participatory budgeting scheme for science will be funded at least initially wholly from general taxation for perpetuity. Ability to own shares in start-ups with this scheme would not be as restrictive as in the case of the Swedish wage-earner funds.

However, George stated that once wage-earner funds became a successful source of finance (as participatory budgets would become), business opposition quickly declined after a heavy initial campaign against the creation of the funds (1993: 476). In the Swedish case, the wage-earner funds were kept separate from Swedish pension funds (Whyman 2007: 236). In order to safeguard the retirement funds of pensioners, their retirement funds would also be kept apart from the investments of the participatory budgeting scheme for science.

The Swedish wage-earner funds were set up on a regional basis too. George argued that this is why the Swedish wage-earner funds were pressured by local unions to take long-term investments into their portfolios rather than engage in speculative investments nationally (1993: 478). Meidner argued that no individual shares would be allocated to workers, unlike the proposed but never implemented Danish wage-earner fund model because allowing workers to sell their shares in firms would have reduced the ability of workers and trade unions to reduce the concentration of ownership over capital. This was because the buyers of the shares would be the existing owners of capital (Meidner 1981: 310). However, George argued that government-appointed boards of union administrators of the Swedish wage-earner funds were not elected or accountable to its stakeholders (1993: 478). Therefore, from the experience of the Swedish wage-earner, the participatory budgeting scheme for science would have to be created separately from pension funds. Funds should be allocated on a regional basis, and individual shares would not be allocated under the scheme.

Many might point to Russia's failed experiment with coupon privatisation as an example of the problems that might occur under a participatory budgeting scheme for science. As part of Russia's coupon privatisation scheme, all citizens of Russia and (then) Czechoslovakia were given a coupon or voucher to buy shares in the industries in which they worked. However, an examination of a number of studies on the subject of Russia's and Czechoslovakia's coupon privatisation by Arup Banerji (1993: 2817), Stephen Fortescue (1994: 137–143), Lynn Nelson and Irina Kuzes (1994: 55–57), Bernard Black, Reiner Kraakman, and Anna Tarassova (2000: 1733), Patrick Flaherty (1992: 4), Andrei Kuznetsov and Olga Kuznetsova (1996: 1174–5) and Stefan Hedlund (2001: 227–31) shows that this example could not be used to say that a participatory budgeting scheme for science would not work. The players involved had the incentive to self-deal for various factors, which happens when transactions are made so that insiders profit at the expense of the company in question and the government does nothing to control this.

The factors behind the self-dealing included the unlikelihood of a long-run rate of return, the previous culture of survival, rent-seeking due to the high degree of policy uncertainty, the lack of confidence in the economic health of Russia, and the inexperience of Russian workers living in a free market economy at the time. The vouchers were made tradeable, and former managers were able to quickly gain monopolised control over their former enterprises because workers did not know about the worth of the coupons. Therefore, there was a quick accumulation of coupons and vouchers in the hands of a few former managers and insiders in Russia in the early 1990s through either intimidation or by buying coupons off unsuspecting workers for extremely low prices or even for bottles of vodka (Banerji 1993: 2821; Nelson and Kuzes 1994: 56–58; Black, Kraakman, and Tarassova 2000:

1733–44; Hedlund 2001: 229–31; Kuznetsov and Kuznetsova 1996: 1181–2; Fortescue 1994: 140–48). The lack of reform of workforces and technology or capitalist culture would not be a problem the participatory budgeting scheme would have to deal with.

The hyperinflation of the early 1990s destroyed the cash savings necessary to facilitate further private investment by (then) Russian citizens. This is something that the participatory budgeting scheme for science would not have to deal with.

Nor would the participatory budgeting scheme for science have to cope with the fiscal policy and institutions of 1990s Russia. Developed economies do not have a punitive tax system nor an unfriendly bureaucracy that increases the incentive for insiders to asset-strip firms (Nelson and Kuzes 1994: 58–60; Banerji 1993: 2822; Fortescue 1994: 136–40; Black, Kraakman, and Tarassova 2000: 1786). These are factors a participatory budgeting scheme in more mature capitalist economies would not have to deal with.

Developed economies have financial laws to protect minority shareholders and workers from being underpaid for shares or from diluting the value of their shares. They also have more experienced lawyers and accountants who advise managers and provide accurate financial disclosure. Therefore, investors are able to know the value of their shares. Securities commissions would be well staffed and have political clout. These are factors that a participatory budgeting scheme in more mature capitalist economies would not have to deal with.

Another issue in 1990s Russia was that their judges and lawyers did not act swiftly and were biased or corrupt (Black, Kraakman, and Tarassova 2000: 1777–95; Kuznetsov and Kuznetsova 1996: 1182; Fortescue 1994: 144–46: Nelson and Kuzes 1994: 59–61; Banerji 1993: 2818–21). These are factors that a participatory budgeting scheme in more mature capitalist economies would not have to deal with.

In summary, the lessons that can be gained from the case of Swedish wage-earner funds is that due to the risky nature of scientific discoveries and commercialisation, the participatory budgeting scheme for science should be kept apart from superannuation or retirement accounts, though after the scheme reaches a certain level of assets and returns, dividends could be paid to pensioners until the end of their lives. Limited funds from the participatory budgeting scheme could be directed to regional areas and to small and medium businesses in order to further build a civil society around the scheme.

In summary, there would be ownership and lending limits for funds invested into start-ups in order to gain returns for taxpayers and also to reduce exposure to financial failure in commercialising technologies. However, these limits would be more lax than under the Swedish wage-earner fund legislation. As previously mentioned, instead of government-appointed union boards, the participatory budgeting scheme for science would hire a board and external experts in science, law, and economics who are not government employees. They would ensure that taxpayers gain the best advice on investments. There would be no tradeable shares or individual holdings under the participatory budgeting scheme for science in order to allow investments to mature and to reduce the administrative workload of updating holding records.

Finally, the funding of the scheme should also be maintained for perpetuity, unlike the Swedish wage-earner funds, which lasted only seven years. For a variety of reasons, the scheme would not face the same challenges that Russia did with its coupon privatisation.

SECTION 3

The State of Science in the World

As mentioned in the introduction, science spending has been subject to variations according to economic cycles, the whims of government, competition against other areas of government spending, and a lack of direct stakeholders in science (as shown in Appendixes 2, 3 and 8). According to the OECD (Van Steen 2012), public investment in science and technology has been hampered throughout the OECD, including Australia, by fiscal constraints (Maass 2003: 43–45). Although this piece has focused on the failure of Australia to build a civil society and culture around the public funding of science and technology, other countries ought to be investing more in science and technology too, especially many middle-income countries, like Poland, Mexico, Indonesia, Hungary, and Turkey, which have equally failed to do so. Since the global financial crisis, there has been a decline in percentages of government spending on research and development as a proportion of GDP across countries such as the United States, Canada, and the United Kingdom.

Countries like Australia should not be complacent about their government funding levels, especially in the face of sustained or rising government spending on research and development by South Korea, Japan, Germany, and China (see Appendix 8). Therefore, a participatory budgeting scheme for science can be relevant and of assistance to these countries too in galvanising public support towards public science spending by allowing the public to directly financially benefit from scientific commercialisation and discoveries.

SECTION 4

Arguments For and Against the Public Funding of Science

Below is a table of existing literature about whether or not to publicly fund science.

Table 1

	Arguments against the public funding of science
Science funding is directed to interest groups and does not serve the public interest.	Breznitz, Dahlman, Dutz, Hodgson, Kaplinsky, Kuznetsov, Lasagabaster, Oldsman, Ornston, Pilat, Sabel, Vijayaraghavan 2014: 36; Smith 1998: 28; Mitcham and Frodeman 2002: 85; Moore 2000: 329.
Interest groups spend more money on gaining new grants than scientific research.	Santamaria, Barge-Gil, and Modrego 2010: 551.
Interest groups who gain grants from the government are wasteful and stifle flow of information to the public.	Foltz 1999: 121; Smith 199: 28; Moore 2000: 329; Mitcham and Frodeman 2002: 89.
Funding is allocated by elites that are accountable to the taxpayer.	Foltz 1999: 117–19.
There is conflict between agencies that are allocated funds.	Santamaria, Barge-Gil and Modrego 2010: 551; Smith and McGeary 1997: 35.

Higher spending did not deliver higher growth as expected.	Foltz 1999: 121.
Science spending led to scandals and higher taxpayer liability for the mistakes of scientists.	Mitcham and Frodeman 2002: 85.
Lower taxes, less regulation would lead to higher science and technology spending as well as better research outcomes.	Hottenrott and Thorwarth 2011: 536; Smith 1998: 34; Armingeon 2007: 317–9; Smith and McGeary 1997: 37; Mitcham and Frodeman 2002: 85. Though some argue that the government's role in the scientific process was only to monitor and evaluate private sector research, provide for necessary science infrastructure such as science parks, streamline the patents process, and fund human skills in science (Breznitz, Dahlman, Dutz, Hodgson, Kaplinsky, Kuznetsov, Lasagabaster, Oldsman, Ornston, Pilat, Sabel, Vijayaraghavan 2014: 19–33).
Public funding of science is too dispersed to be effective.	Hicks and Katz 2011 137–151; Hunter 2013: 1047-9.
Government departments do not follow price signals and budget constraints.	Smith 1998: 28–29.
Private sector has better market information than government departments.	Breznitz, Dahlman, Dutz, Hodgson, Kaplinsky, Kuznetsov, Lasagabaster, Oldsman, Ornston, Pilat, Sabel, Vijayaraghavan 2014: 20.

Adrian Soh

Below is a table of all the literature citing arguments for the public funding of science and solutions to the problems of the public funding of science.

Table 2

	Arguments for the public funding of science
Science funding is directed to interest groups and does not serve the public interest.	Government allocation of science funding is influenced by the demands of the public (Moore 2000: 329). Although Moore states that it is too focused on the demands of the public, there could be a change in the criteria by which government grants are allocated. Criteria could include the number of collaborating research agencies and businesses in one project, product approvals, and the quantity and quality of research training programmes of new research workforces. Grants should also be allocated according to the earnings of employees, clinical trial data, scientific equipment purchased, and the diversity of employees on research projects, such as the number of women employed in projects, according to Lane, who also stated that the data above ought to be included in the proposals and the reports on projects to the government (2014: 7–13). Other criteria could also include figures such as the estimated number of businesses that could use a new technology, the estimated amount of resources that could be saved, and the estimated number of patients that could be treated in the case of medical innovations, or even if the research could evolved or further other areas of research. Research in different geographical locations and ensuring that the research is not being undertaken anywhere else in the country ought to be requirements for grant consideration under the participatory budgeting scheme for science. Furthermore, legal attachments could be made to lessen patent periods of research or to set fees to be paid to attain the results of research that is funded by the participatory budgeting scheme for science.

Interest groups spend more money on gaining new grants than on scientific research.	The shortage of public funding for science led scientists to spend time lobbying the private sector for short-term funding rather than concentrating on groundbreaking research (Hottenrott and Thorwarth 2011: 535–6). The lack of public funding for science research led them to spend time lobbying for greater government grants (Smith and McGeary 1997: 36).
Interest groups who gain grants from the government are wasteful and stifle flow of information to the public.	The greater dependence of universities on private sector grants stifled information and caused bottlenecks in scientific research because of confidentiality agreements with the private sector. Overuse of private sector investment could also lead to the private sector dictating university research. This is known as the "skewing problem" (Hottenrott and Thorwarth 2011: 535–37; Howard and Laird 2013: 72; Maass 2003: 47–50). Science research key performance indicators, such as the number of citations by scientific peers in good quality academic journals, could be created; research could become more contract based; or by other performance criteria as suggested above (Hicks and Katz 2011: 148–49; Gramm, Lieberman, Domenici, and Bingaman 1998: 21; Maass 2003: 45–46). Government spending on science helped fuel growth in private sector investment in science, and it attracts foreign human capital (Moore 2000: 330; Mitcham and Frodeman 2002: 86). As seen in section 1, increased public science spending also leads to increased productivity and the creation of new industries.
Funding is allocated by elites that are accountable to the taxpayer.	Government allocation of science funding is influenced by the demands of the public (Moore 2000: 329). However, Moore states that it is too focused on the demands of the public.

There is conflict between agencies that are allocated funds.	Funding constraints have led to increased collaboration between public and private research institutions (Smith and McGeary 1997: 36). As suggested before, government grant criteria could also mandate cooperation between scientific institutions, including different public sector institutions.
Higher spending did not deliver higher growth as expected.	This is because the revenues of scientific discoveries that could be commercialised were not distributed to a wider public, as the participatory budgeting scheme for science proposes. The more equal distribution of wealth from scientific discoveries and the financing of commercialised technologies that otherwise would not have gained finance have a better likelihood of spurring greater economic growth.
Science spending led to scandals and higher taxpayer liability for the mistakes of scientists.	The purchasing of insurance can be added to any loan or investment given, especially in the case of small and medium start-ups.
Lower taxes and less regulation would lead to higher science and technology spending as well as better research outcomes.	The increase in scientific patents has led to rent-seeking behaviour, when companies use their monopoly ownership of patents to charge high rents to use their technology and to stifle competition. Increases in patents also have led to more resources being directed to litigation costs (Boldrin and Levine 2013: 4–5). There is nothing stopping governments from using these policy instruments alongside increases in public funding of science.
Public funding of science is too dispersed to be effective.	A balanced portfolio of science funding is important because certain areas of scientific research are dependent on others. Therefore, focusing on one area of scientific research over others leads to bottlenecks in scientific research as a whole (Mitcham and Frodeman 2002: 85; Smith and McGeary 1997: 37).

Government departments do not follow price signals and budget constraints.	The OECD stated that governments tried to deal with the shortage of funding through competitive grant processes (Van Steen 2012: 6–8). Funding constraints have led to increased collaboration between public and private research institutions (Smith and McGeary 1997: 36). Budget constraints led to more selective funding of research projects within institutions (Robins 1980: 129–33).
The private sector has better market information than government departments.	Governments can hire private sector consultants, or they can collect information through existing government departments or by engaging stakeholders. Information is never perfect; therefore governments need to step in when the private sector will not due to the uncertainty of the return on investment (Greenstein 2014: 94–95; Howard and Laird 2013: 74; Santamaria, Barge-Gil, and Modrego 2010: 549–50).

Therefore, in order to address the problems with the public funding of science with the participatory budgeting scheme for science will have to be implemented alongside measures to reduce risk of financial failures and to direct more funds to small and medium start-ups, as seen in section 2. The suggested criteria above—hiring external consultants for advice; greater stakeholder engagement; and the board of the fund being made independent of government departments—would help reduce the diversion of funds to special interests and lead to a more balanced research portfolio.

All these measures would also allow market information about emerging technological areas to be used in the investment decisions of the participatory budgeting scheme for science. The scheme would maintain budget constraints on science but allow for private sector involvement and investment. This would still force private and public research institutes to collaborate and to reduce budget constraints on research. Meanwhile, the gradual increase in dividends for pure research once the fund's assets reach a certain level will gradually reduce the need for scientific researchers to spend time seeking further grants from the public and private sectors. Grants would allow for public and private cooperation in research. Clauses in agreements regulating the royalties from research might allow for better access to information and the building of further knowledge.

In section 1, we saw that the distribution of dividends to a large number of people would contribute to increased consumption from built-up assets and returns earned from providing finance to technology start-ups. The participatory budgeting scheme for science would not mean that reduced taxes on science research, tax incentives for commercialised research, and the deregulation of research would not be attempted alongside the scheme. When the fund starts giving distributions to pure research, proposals can include the purchase of insurance to protect taxpayers from legal liability from faultily conducted research.

SECTION 5

Areas of Further Research, and Steps to Implementation

In conclusion, a participatory budgeting scheme for science funding is necessary and is workable under certain conditions, such as investment limits and risk management strategies. It is necessary to ensure that we allow for a greater proportion of the population to share in the wealth from new industries that will be created in the twenty-first century and that we allow for more sustained consumption and investment in economies.

Furthermore, such a scheme will build a civil society around science funding that will drive productivity and competitiveness and will deal with market failure to invest in science and technology due to the uncertainty of return. Renewed grant selection criteria, more stakeholder involvement, engagement in how science spending is dispersed, and making the funding of science and technology more constant and harder to undo will allow for a more balanced allocation of science funding. It will alleviate the lack of infrastructure and the human resource bottlenecks that hinder groundbreaking scientific discoveries.

Furthermore, it will assist in the de-politicisation of science and technology spending and will address inconsistencies in funding as the assets and returns of the scheme increase. The de-politicisation of science funding will likewise be helped by hiring an expert board along with external consultants, persons who are not employees of the government and who have specialized qualifications.

Small and medium start-up businesses and universities, including scientific institutions, will likewise benefit from the scheme because their research will attract finance, lift their financial burdens, and allow them to take advantage of increased commercial revenue, albeit indirectly. Cooperation between stakeholders will be crucial as governments battle with budget constraints. New grant criteria should encourage more of this.

A participatory budgeting scheme would also address the problem of a lack of information sharing within the research community. The different treatment of pure and commercialised research is antithetical to any participatory budgeting scheme for science. Furthermore, grant applications, loans, and equity purchases have to take into account whether or not foreign knowledge can be acquired.

A participatory budgeting scheme for science would certainly be successful as many Western nations have the institutions and culture to facilitate its successful implementation, which post-Soviet Russia and former Czechoslovakia did not have. Finally, Western nations do not have to deal with the challenges of macroeconomic instability created by years of economic control.

SECTION 6

Further Government Action and Areas to Research

The funds used in the participatory budgeting scheme for science should not be placed in companies listed on the stock exchange. The programme should be perpetual, and the limits on ownership should be less than those seen in the case of the Swedish wage-earner funds. The shares should be held in trust by the government, funds should be kept separate from retirement funds, and surplus funds should directly supplement the income of citizens and pure research grants. Super-profit taxes certainly could fund such a scheme, but reduced tax breaks or redirecting current corporate subsidies could also fund the scheme.

In the future, governments will need to introduce the scheme on a pilot basis, at least initially, to ensure that they learn from errors. Significant stakeholder engagement must take place in order to settle any concerns with the structure of the scheme and the generation of a new science grant scheme. Stakeholder engagement must take place to gather information about the future developments of science in which investment should be made.

Governments ought to examine incentives to take up science and mathematics studies and careers and should ensure that science and mathematics subjects are taught effectively. They need to streamline and shorten time frames for patents on technology and streamline taxes on technology to boost science and technology spending alongside building a participatory budgeting scheme for science.

Governments should retain a portion of the science budget in its direct control to fund pure research through a reformed grants process. They should allow the participatory budgeting for science scheme to focus on the commercialisation of technology by granting loans and purchasing equity in start-up firms until the scheme has sufficient assets to grant funds for pure research and distributions to taxpayers. The process would also require external consultants in finance and science would to assist in the reduction of asymmetric information about possible future growth areas and financial returns to the scheme.

The criteria for grants and the desired skill sets of consultants and scheme employees need to be researched further. How to fund the participatory budgeting scheme for science and how to conduct a proper cost-and-benefit analysis of the scheme could likewise be questions for research. Further research is needed regarding what affect the scheme would have on private donations to scientific research.

APPENDIX 1

Grouping of Countries in the OECD According to Percentage of Research and Development Spending Levels

Range of Percentage of Research and Development	Countries Within This Range
4 to 4.5 per cent of gross domestic product (GDP)	Korea and Israel
4 per cent to 3 per cent of GDP	Japan, Finland, Sweden, Denmark, and Austria
3 per cent to 2 per cent of GDP	Germany, United States, Slovenia, OECD average, Belgium, France, Australia, China, and Netherlands
2 per cent to 1 per cent of GDP	Czech Republic, Estonia, Norway, Great Britain, Canada, Ireland, Hungary, Italy, Spain, New Zealand, Luxembourg, and Russia
1 per cent to 0 per cent of GDP	Turkey, Poland, Slovakia, Greece, and Mexico

Source: OECD Science, Technology and Industry Scoreboard 2015

APPENDIX 2

Australian Nominal and Real Government Spending on Science and Technology 2006–2016

Source: Australian Federal Budget Papers 2015–2016	Nominal spending on science and technology by millions of dollars	Approximate real spending on science and technology by millions of dollars (AUD) at March 2006 prices
2006–2007	6,613.4	6613.4
2007–2008	6,718.7	6555.7
2008–2009	7,515.0	7032.3
2009–2010	8,456.2	7724.8
2010–2011	8,963.7	7956.2
2011–2012	10,109.4	8690.1
2012–2013	9,547.4	8075.6
2013–2014	10,085.4	8322.4
2014–2015	10,032.7	8043.2
2015–2016 est.	9,717.0	7688.0

Source: Australian Federal Budget Papers 2015-2016

APPENDIX 3

Changes in Funding Sources as a Percentage of GDP since 2000 to 2014 in the OECD

	Government funded research and development as a percentage of GDP across the OECD	Industry funded research and development as a percentage of GDP across the OECD	Industry financing as a percentage of total research and development spending across the OECD	Government spending as a percentage of total research and development spending across the OECD
2000	0.61	1.37	64.05	28.48
2001	0.63	1.38	63.40	28.91
2002	0.64	1.34	62.06	29.84
2003	0.65	1.33	61.57	30.33
2004	0.65	1.31	61.55	30.49
2005	0.64	1.34	62.20	29.64
2006	0.63	1.38	63.19	28.82
2007	0.63	1.41	63.44	28.47
2008	0.67	1.44	62.63	29.26
2009	0.73	1.38	58.94	31.22
2010	0.72	1.35	58.61	31.12
2011	0.69	1.40	59.88	29.74
2012	0.68	1.40	60.09	29.22
2013	0.67	1.44	60.58	28.35
2014		1.44	60.92	

Source: OECD

APPENDIX 4

Comparative Gross Expenditures on Research and Development (by Millions and Current PPP $)

Year	AUS	CAN	FIN	GER	JAP	KOR	UK	US	CHINA	ISR
2000	7964.09	16746.65	4446.343	52361.42	98758.01	18533.1	27865.34	269513	33044.55	6162.78
2001		18967.71	4569.478	54466.58	103825.7	21275.01	29200.85	280238	38547.69	6725.76
2002	9885.30	19145.33	4814.678	56657.03	108166.2	22506.78	30635.69	279891	48059.9	6845.65
2003		20133.52	4960.435	59533.38	112192.3	24071.71	31096.71	293852	57137.22	6207.44
2004	11683.52	21643.01	5387.241	61314.43	117597.9	27942.35	32015.81	305640	70131.7	6662.09
2005		23089.97	5601.228	64298.79	128694.6	30618.33	34080.66	328128	86827.63	6966.3
2006	15509.04	24091.54	6064.391	70185.38	138564.9	35413.07	37022.73	353328	105580.6	7503.08
2007		24742	6636.473	74015.66	147602.2	40640.27	38731.05	380316	124187.1	8728.38
2008	19133.00	24911.9	7487.877	81970.66	148719.2	43906.41	39396.93	407238	146126.6	8706.36
2009		25046.83	7512.354	82795.64	136954	45987.25	39420.23	406405	185266.7	8501.20
2010	20572.19	25048.23	7658.35	87882.64	140607.4	52172.78	38165.61	410093	213460.1	8658.73
2011	20955.60	25674.58	7892.045	96282.45	148389.2	58379.65	39132.64	428745	247808.3	9523.41
2012		26278.98	7486.207	100697.1	152325.6	64862.49	38811.94	436078	292062.9	10448.79
2013	23083.98	26303.76	7321.691	102573	162347.2	68051.5	41743.39	456977	333521.6	10998.93
2014		25813.56	7050.834	106780.8	166861.3	72266.75	44174.09		368731.6	11376.5

Source : OECD

AUS - Australia, CAN - Canada, FIN - Finland, GER - Germany, JAP - Japan, KOR - Korea, UK - United Kingdom, US - United States, ISR - Israel

APPENDIX 5

Comparative Compound Growth in Gross Expenditure on Research and Development (%)

	AUS	CAN	FIN	GER	JAP	KOR	UK	US	CHINA	ISR
2000	4.29	11.78	12.19	5.51	3.09	14.91	2.30	7.32		
2001		10.74	1.07	1.44	2.79	12.24	2.13	1.66	13.41	6.70
2002	9.35	0.51	3.57	1.25	1.63	4.35	2.58	-1.63	22.78	-1.34
2003		1.52	3.39	0.98	2.61	6.45	0.74		16.57	-4.51
2004	6.23	4.62	4.33	-0.30	2.01	12.97	-1.11	1.23	19.47	4.45
2005		1.77	3.24	0.78	6.98	7.77	4.04	4.01	19.90	8.82
2006	11.20	1.03	4.30	5.13	4.78	13.37	3.92	4.47	17.97	8.13
2007		0.04	5.44	2.85	3.76		4.73	4.85	14.58	13.31
2008	8.84	-1.44	6.78	7.31		7.05	-0.27	5.02	15.40	1.23
2009		0.12	-3.05	-1.01	-8.52	6.18	-1.16	-0.96		-3.60
2010	0.92	-1.20	2.36	3.59	1.43	12.08	-1.15	-0.31	13.82	0.62
2011	0.52	0.77	0.17	6.80	3.50	11.99	1.73	2.43	13.75	7.20
2012			-7.37	3.23	0.55	10.00	-2.94	-0.13	15.78	5.90
2013	2.18	-3.58	-4.67	-1.28		6.04	4.84	3.11	12.52	2.24
2014		-2.22	-4.12	2.16	3.04	6.87	5.19		8.97	3.07

Source: OECD
AUS – Australia, CAN – Canada, FIN – Finland, GER – Germany, JAP – Japan, KOR – Korea, UK – United Kingdom, US – United States, ISR – Israel

APPENDIX 6

Comparative Researchers Per Thousand People in the Labour Force

	AUS	CAN	FIN	GER	JAP	KOR	UK	US	CHINA
2000	6.91	6.787	13.4	6.52	9.57	4.90	5.93	6.83	0.94
2001		7.092	14	6.66	9.67	6.07	6.33	6.98	1.01
2002	7.417	6.979	14.7	6.71	9.31	6.19	6.83	7.16	1.09
2003		7.249	15.9	6.81	9.79	6.59	7.41	7.61	1.15
2004	7.995	7.579	15.7	6.76	9.84	6.67	7.69	7.42	1.23
2005		7.878	15	6.65	10.23	7.57	8.27	7.31	1.47
2006	8.162	8.009	15.1	6.75	10.28	8.34	8.31	7.40	1.60
2007		8.45	14.5	6.99	10.24	9.16	8.22	7.34	1.86
2008	8.229	8.644	15	7.26	9.84	9.70	8.10	7.65	2.07
2009		8.202	15.1	7.61	9.86	10.01	8.21	8.04	1.49
2010	8.592	8.568	15.4	7.87	9.89	10.67	8.18	7.71	1.54
2011		8.836	14.8	8.22	9.96	11.51	7.95	8.08	1.68
2012		8.561	14.9	8.53	9.86	12.38	8.02	8.09	1.78
2013		8.333	14.5	8.50	10.04	12.44	8.32		1.87
2014			14.2	8.57	10.37	13.02	8.44		1.91

Source: OECD

AUS – Australia, CAN – Canada, FIN – Finland, GER – Germany, JAP – Japan, KOR – Korea, UK – United Kingdom, US – United States, ISR – Israel

APPENDIX 7

Full-Time Equivalents of Researchers in Selected Countries

	AUS	CAN	FIN	GER	JAP	KOR	UK	US	CHINA
2000	66001	107900	34847	257874	647572	108370	170554	983208	695062
2001		114510	36889	264385	653021	136337	182144	1013307	742726
2002	73173	115960	38630	265812	623035	141917	198163	1047242	810525
2003		123230	41724	268942	652369	151254	216690	1126251	862108
2004	81192	130380	41004	270215	653747	156220	228969	1105097	926252
2005		136700	39582	272148	680631	179812	248599	1101062	1118698
2006	87201	140660	40411	279822	684884	199990	254009	1130182	1223756
2007		151330	39000	290853	684311	221928	252651	1133557	1423381
2008	92649	157200	40879	302641	656676	236137	251932	1191024	1592420
2009		150220	40849	317307	655530	244077	256124	1250984	1152311
2010	100414	158660	41425	327996	656032	264118	256585	1198280	1210841
2011		165100	40003	338689	656651	288901	251358	1252948	1318086
2012		161590	40468	352419	646347	315589	256156	1265064	1404017
2013		159190	39196	354463	660489	321842	267699		1484040
2014			38281	359640	682935	345463	273560		1524280

Source: OECD
AUS - Australia, CAN - Canada, FIN - Finland, GER - Germany, JAP - Japan, KOR - Korea, UK - United Kingdom, US - United States, ISR - Israel

APPENDIX 8

Comparative Data on Government Spending on Research and Development by Select Countries

	AUS	CAN	FIN	GER	JAP	KOR	UK	US	CHINA	ISR
2000	0.67	0.55	0.85	0.75	0.59	0.52	0.52	0.69	0.30	0.86
2001		0.59	0.82	0.75	0.58	0.58	0.49	0.73		0.84
2002	0.68	0.63	0.85	0.76	0.57	0.58	0.50	0.76		0.80
2003		0.62	0.85	0.77	0.57	0.56	0.53	0.78	0.34	0.79
2004	0.70	0.62	0.87	0.74	0.57	0.59	0.53	0.79	0.33	0.68
2005		0.63	0.85	0.69	0.55	0.60	0.53	0.77	0.35	0.58
2006	0.75	0.61	0.84	0.68	0.55	0.65	0.53	0.76	0.34	0.55
2007		0.61	0.80	0.67	0.54	0.74	0.52	0.77	0.34	0.54
2008	0.78	0.64	0.77	0.74	0.54	0.79	0.52	0.84	0.34	0.53
2009		0.66	0.90	0.81	0.59	0.90	0.57	0.92	0.39	0.53
2010		0.65	0.96	0.82	0.56	0.93	0.55	0.89	0.41	0.56
2011		0.61	0.91	0.83	0.56	0.93	0.51	0.86	0.39	0.52
2012		0.61	0.91	0.84	0.56	0.96	0.47	0.80	0.42	0.53
2013		0.58	0.86	0.82	0.60	0.95	0.48	0.76	0.43	0.52
2014		0.56	0.87		0.57	0.99	0.49		0.41	

Source: OECD

AUS – Australia, CAN – Canada, FIN – Finland, GER – Germany, JAP – Japan, KOR – Korea, UK – United Kingdom, US – United States, ISR – Israel

REFERENCES

Armingeon, K. (2007), "Two Perspectives on EU Research Funding: The Present Is Lacklustre, The Future Is Potentially Shining", *European Political Science*, vol. 6, 315–21.

Asian Development Bank (2014), "ADB-OECD Study on Enhancing Financial Accessibility for SMEs Lessons from Recent Crises", Asian Development Bank, Manila.

Australian Clean Energy Finance Corporation (2014), "Annual Report 2013–2014", http://www.cleanenergyfinancecorp.com.au/reports/annual-reports/files/annual-report-2013-14.aspx.

Australian Taxation Office (2015), "Taxation Statistics", https://data.gov.au/dataset/taxation-statistics-2012-13/resource/233cbf28-6fda-4e53-bbe9-3a37a65fb742.

Australian Treasury Department (2010), "Australia's Future Tax System Report to the Treasurer", *Australian Treasury Department*, Canberra.

—— (2015), "The Australian Budget 2015–2016", *Australian Treasury Department*, Canberra.

Banerji, A. (1993), "Privatisation in the Russian Federation", *Economic and Political Weekly*, vol. 28, no. 51, 2817–22.

Black, B., R. Kraakman, and A. Tarassova (2000), "Russian Privatization and Corporate Governance: What Went Wrong?", *Stanford Law Review*, vol. 52, no. 6, 1731–1808.

Boldrin, M., and D. Levine (2013), "The Case Against Patents, Journal of Economic Perspectives", vol. 27, no. 1, 3–22.

Breznitz, D., C. Dahlman, M. A. Dutz, B. Hodgson, R. Kaplinsky, Y. Kuznetsov, E. Lasagabaster, E. Oldsman, D. Ornston, D. Pilat, C. Sabel,

K. Vijayaraghavan (2014), "Making Innovation Policy Work: Learning From Experimentation", OECD, Paris.

Caner, M. and T. Grennes (2010), "Sovereign Wealth Funds: The Norwegian Experience", *The World Economy*, vol. 33, no. 4, 597–614.

Centre for International Economics (2015), "The Importance of Advanced Physical and Mathematical Sciences to the Australian Economy", Australian Academy of Sciences, Canberra.

Chubb, I., C. Findlay, L. Du, B. Burmester, L. Kusa (2012), "Mathematics, Engineering and Science in the National Interest", Australian Government: Office of the Chief Scientist, 6–11.

Clean Energy Finance Corporation (2012), "Clean Energy Finance Corporation Act 2012", http://www.cleanenergyfinancecorp.com.au/about/governance.aspx.

Flaherty, P. (1992), "Privatisation and the Soviet Economy", *Monthly Review*, vol. 43, no. 8, http://archive.monthlyreview.org/index.php/mr/article/view/MR-043-08-1992-01_1.

Foltz, F. (1999), "Five Arguments for Increasing Public Participation in Making Science Policy", *Bulletin of Science Technology and Society*, vol. 19, 117–27.

Fortescue, S. (1994), "Privatisation of Russian Industry", *Australian Journal of Political Science*, vol. 29, 135–54.

George, D. (1993), "The Political Economy of Wage-Earner Funds: Policy Debate and Swedish Experience", *Review of Political Economy*, vol. 5, no. 4, 470–90.

Gramm, P., J. Lieberman, P. Domenici, and J. Bingaman (1998), "The Long Road to Increase Science Funding", *Issues in Science and Technology*, vol. 14, no. 3, 21.

Greenstein, S. (2014), "The Irony of Public Funding", *IEEE Micro*, vol. 34, no. 1, 94–95.

Hedlund, S. (2001), "Property Without Rights: Dimensions of Russian Privatisation", *Europe-Asia Studies*, vol. 53, no. 2, 213–237.

Hicks, D., and J. S. Katz (2011), "Equity and Excellence in Research Funding", *Minerva*, vol. 49, 137–151.

Hottenrott, H., and S. Thorwarth (2011), "Industry Funding of University Research and Scientific Productivity", *KYKLOS*, vol. 64, no. 4, 534–55.

Howard, D. J., and F. N. Laird (2013), "The New Normal in Funding University Science", *Issues in Science and Technology*, 71–76.

Hunter, P. (2013), "EU-Life Revives Funding Debate", *European Molecular Biology Organisation Reports*, vol. 14, no. 12, 1047–9.

Kut, C., B. Pramborg, and J. Smolarski (2007), "Managing Financial Risk and Uncertainty: The Case of Venture Capital and Buy-Out Funds", *Global Business and Organizational Excellence*, vol. 26, no. 2, 53–64.

Kuznetsov, A., O. Kuznetsova (1996), "Privatisation, Shareholding and the Efficiency Argument: Russian Experience", *Europe-Asia Studies*, vol. 48, no. 7, 1173–85.

Lamberts, R., W. J. Grant, and A. Martin (2010), "ANU Poll Public Opinion about Science", *Australian National University*, 5–24.

Lane, J. (2014), "New Linked Data on Research Investments: Scientific Workforce, Productivity, and Public Value", *American Institutes for Research*, Washington.

Maass, G. (2003), "Funding of Public Research and Development: Trends and Changes", *OECD Journal on Budgeting*, vol. 3, no. 4, 59.

Meidner, R. (1981), "Collective Asset Formation through Wage-Earner Funds", *International Labour Review*, vol. 120, no. 3, 303–16.

Mitcham, C., and Frodeman, R., 2002, The Plea for Balance in the Public Funding of Science, Technology in Science, vol. 24, 83–92.

Moore, A. (2000), "Science Funding and Infrastructures In Europe", *TIG*, vol. 16, no. 8, 329–30.

Nelson, L., and I. Y. Kuzes (1994), "Evaluating the Russian Voucher Privatization Program", *Comparative Economic Studies*, vol. 36, no. 1, 55–67.

OECD (2013), "Supporting Investment in Knowledge Capital, Growth and Innovation", *OECD*, Paris.

OECD (2015), "New Approaches to SME and Entrepreneurship Financing: Broadening the Range of Instruments", *OECD*, Paris.

Pontusson, J., and S. Kuruvilla (1992), "Swedish Wage-Earner Funds: An Experiment in Economic Democracy", *Industrial and Labour Relations Review*, vol 45, no. 4, 779–91.

Rehn, G. (1957), "Swedish Wages and Wage Policies", *Annals of the American Academy of Political and Social Science*, vol. 310, 99–108.

Rehn, G., 1957, Swedish Wages and Wage Policies, Annals of the American Academy of Political and Social Science, vol. 310, 99–108.Robins, M. O. (1980), "Methods of Funding National Research", *Physical Technology*, vol. 11, 128–33.

Santamaria, L., A. Barge-Gil, and A. Modrego (2010), "Public Selection and Financing of R&D Cooperative Projects: Credit Versus Subsidy Funding", *Research Policy*, vol. 39, 549–63.

Smith, F. (1998), "Government Research Funding and Economic Distortion", *Knowledge, Technology and Policy*, vol. 11, no. 3, 27–39.

Smith, P. M., and M. McGeary (1997), "Don't Look Back: Science Funding for the Future", *Issues in Science and Technology*, vol. 13, no. 3, 33–40.

Van Steen, J. (2012), "Modes of Public Funding of Research and Development: Towards Internationally Comparable Indicators", *OECD Science, Technology and Industry Working Papers*, no. 4, OECD, Paris.

White, S. (2014), "Ownership for All: The Lost Radicalism of the Centre".

White, S. and N. Seth-Smith, eds. (2014) "Democratic Wealth: Building a Citizens' Economy", http://www.opendemocracy.net/ourkingdom/opendemocracy/democratic-wealth-free-e-book.

Whyman, P. (2006) "Post-Keynsesianism, Socialization of Investment and Swedish Wage-Earner Funds", Cambridge Journal of Economics, vol. 30. 49–68.

Printed in the United States
By Bookmasters